Lah'aj Kids Addition Worksheets
With Ṡabaṫiyya

By Tiyi Hibner
©2025 Yamartat Ta'/Tiyi Hibner
Published by www.TempleBabies.com

ṠABAṪ•IYYA HARAF•AAT SABAEIC ALPHABET

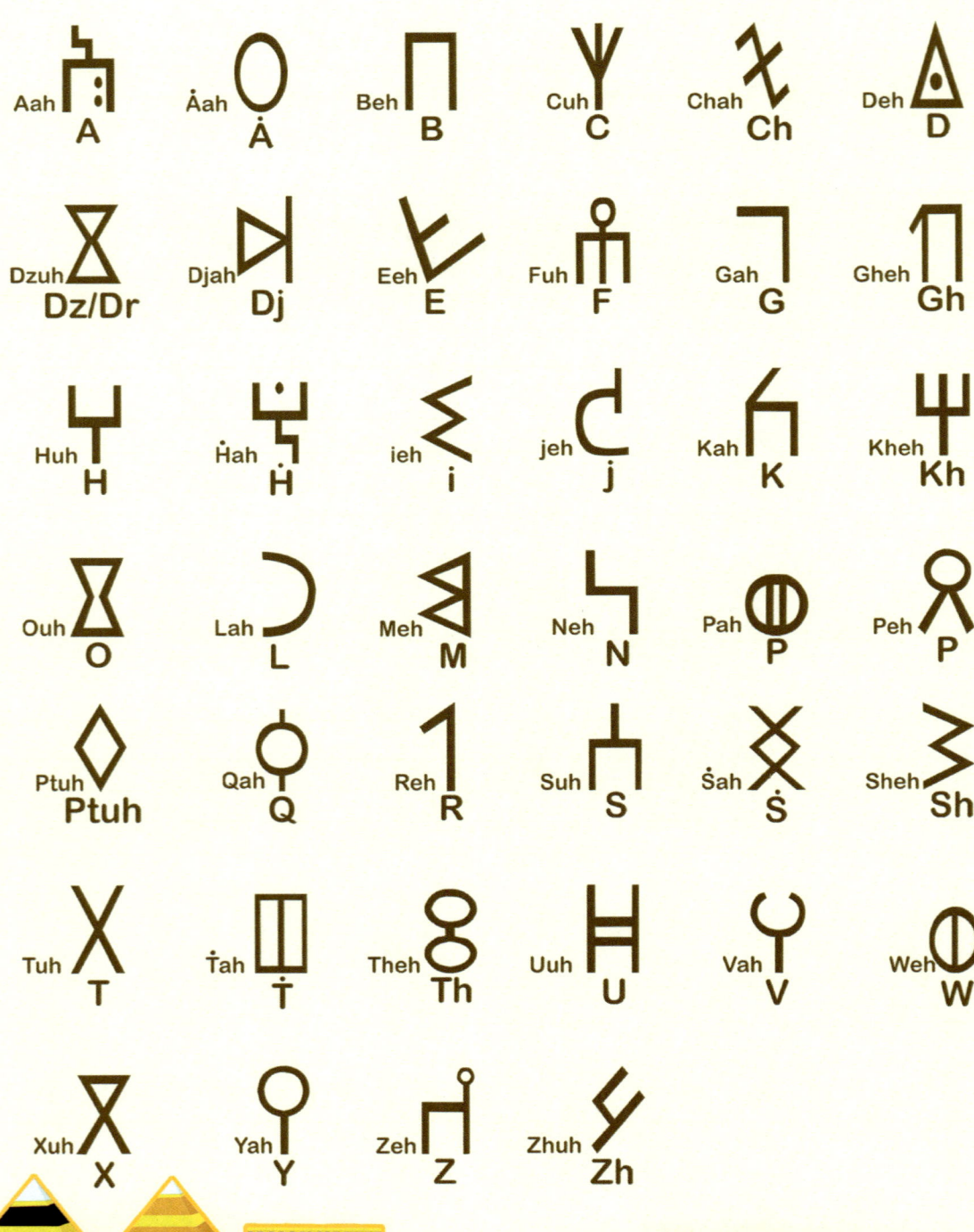

Ẋ•⊡•⊡ • ⊡•⊡•1 ⊡•⊡ • ⊡•⊡•Ẋ
ṠABAṪ•IYYA RAQAM•AAT SABAEIC NUMBERS

⊙ 0 Safu	I 1 Wahu	⋀ 2 Athu	⅄ 3 Thalu	+ 4 Rabu	✕ 5 Khamu
✳ 6 Satu	✳ 7 Sabu	✳ 8 Tamu	⊕ 9 Tasu	∩ 10 Ashu	∩I 11 Ashu Wu Wahu
∩⋀ 12 Ashu Wu Athu	∩⅄ 13 Ashu Wu Thalu	∩+ 14 Ashu Wu Rabu	∩✕ 15 Ashu Wu Khamu
∩✳ 16 Ashu Wu Satu	∩✳ 17 Ashu Wu Sabu	∩✳ 18 Ashu Wu Tamu	∩⊕ 19 Ashu Wu Tasu
∩∩ 20 Athura	∩∩I 21 Athura Wahu	∩∩⋀ 22 Athura Athu	∩∩⅄ 23 Athura Thalu
∩∩+ 24 Athura Rabu	∩∩✕ 25 Athura Khamu	∩∩✳ 26 Athura Satu	∩∩✳ 27 Athura Sabu
∩∩✳ 28 Athura Tamu	∩∩⊕ 29 Athura Tasu	30 Thalura	40 Rabura
50 Khamura	60 Satura	70 Sabura	80 Tamura

90 Tasura	Mayu 100	Mayu Mayu 200	Afu 1,000	Ashu Afu 10,000	Mayu Afu 100,000	Malu 1,000,000

/∩∩✕·25

᚛ᚐᚁᚐᚅ᚜·RANAN·NAME

᚛ᚐᚁᚐᚒᚈ᚜·KHADUT·DATE

Lah'aj Kids

ADDITION WORKSHEETS WITH ṠABAṬIYYA

ᚦᚨᚾᚨᚾ · RANAN · NAME

Хᚻᚨᚾᚥ · KHADUT · DATE

∩∧ + ∩	✻ + +	∩∧ + ∧	∧ + ∧	∩ + ⊛
+ ∧∩	+ ∪✻	⊛ + ∩∐	✻ + ∧	✻ + +
+ ✻	+ ✻	+ ✻	∧ + ∩	∐✻
⊛ + ∩∧	∩∐	⊛ + ∧	∩∐ + ∐	✻∐ + ∩
+ ⊛	✻ + ⊛	∩∧ + ✻	∐∐ + ∩∧	∧ + +

ᛁᚾᛁᛌ·RANAN·NAME

ᚷᚻᚫᛙᛙ·KHADUT·DATE

+ ✳✳

+ ∩∧ / ∩∧

+ ∩⊒

+ ✕✕

+ ✳ / ⊛

+ ⊛ / ✕

+ ⅄ / ✳

+ ✕ / ✕

+ ✳ / ∩∧

+ ∧ / ✕

+ ✳✳

+ ✳ / ✕

+ ⅄ / ⊛

+ ∩ / ✳

+ ∧ / ∩∧

+ ∩∧ / ⊛

+ ∩⊒ / ⊛

+ ᛭ / ✕

+ ∩∧ / ✕

+ ✳✳

+ ⊛ / ✕

+ ✳✳

+ ⅄ / ∩⊒

+ ∩ / ✕

+ ∩∧ / ✳✳

ꓤ꒦ꓤ꒦ꓢ · RANAN · NAME

/∩∩ᚷ · 25

ᚷᚻᨔᚻᛈ · KHADUT · DATE

/ⴖⴖ⋇ · 25

᭬ᔆᔆᔆᔆ · RANAN · NAME

ᚷᚻᚥᔆᚥ · KHADUT · DATE

/∩∩✳ · 25

ᔕᕼᕼᔕ᐀ · RANAN · NAME

ᕽᕼᐃᕼᕼᒎ · KHADUT · DATE

Lak'aj Kids

ADDITION WORKSHEETS WITH ŚABATIYYA

ᔔᔓᔔᔕ1·RANAN·NAME

ᕝᑎᐃᔕᙡ·KHADUT·DATE

Lah'aj Kids

/⊓⊓✕·25

ᚷᚨᚾᚨᚾ·RANAN·NAME

ᚷᚺᚪᚾᚢᛏ·KHADUT·DATE

Lak'aj Kids

/ᗰᗰ✕ · 25

ᒣᗩᒣᗩᒧ · RANAN · NAME

ᕽᕼᗩᕽᙀ · KHADUT · DATE

$$+\begin{array}{r}+\\+\end{array}$$
$$+\begin{array}{r}+\\*\end{array}$$
$$+\begin{array}{r}⋇\\∪∨\end{array}$$
$$+\begin{array}{r}∩⊓\\*\end{array}$$
$$+\begin{array}{r}⋇\\∩\end{array}$$

$$+\begin{array}{r}⅄\\*\end{array}$$
$$+\begin{array}{r}*\\*\end{array}$$
$$+\begin{array}{r}∩*\\∪\end{array}$$
$$+\begin{array}{r}⋇\\⅄\end{array}$$
$$+\begin{array}{r}∩⊓\\⋇*\end{array}$$

$$+\begin{array}{r}∪\\∨\end{array}$$
$$+\begin{array}{r}⅄\\—\end{array}$$
$$+\begin{array}{r}∪*\\∪∨\end{array}$$
$$+\begin{array}{r}⋇⋇\\*\end{array}$$
$$+\begin{array}{r}+\\⋇*\end{array}$$

$$+\begin{array}{r}+\\◉\end{array}$$
$$+\begin{array}{r}*\\*\end{array}$$
$$+\begin{array}{r}*\\—\end{array}$$
$$+\begin{array}{r}*\\∨\end{array}$$
$$+\begin{array}{r}*\\*\end{array}$$

$$+\begin{array}{r}—\\*\end{array}$$
$$+\begin{array}{r}⋇\\⋇\end{array}$$
$$+\begin{array}{r}∩⊓\\∪\end{array}$$
$$+\begin{array}{r}—\\∩\end{array}$$
$$+\begin{array}{r}—\\∩⊓\end{array}$$

_____ ·RANAN·NAME

_____ ·KHADUT·DATE

ꓵ∧
+ ꓵ∧
_____

⊙
+ ⚹
_____

∧
+ ⚹
_____

⚹
+ ⚹
_____

⊛
+ |
_____

⅄
+ +
_____

⊙
+ |
_____

⚹
+ ⚹
_____

ꓵ
+ ⊙
_____

∧
+ ⊙
_____

ꓵ∧
+ +
_____

ꓵ⊔
+ ⚹
_____

⊙
+ ⚹
_____

⚹
+ ꓵ∧
_____

+
+ ⚹
_____

⚹
+ ꓵ
_____

|
+ +
_____

⚹
+ ⚹⚹
_____

∧
+ ⊛
_____

⚹
+ ⅄
_____

ꓵ
+ ⊛
_____

⚹
+ ⚹
_____

⊛
+ ⊛
_____

+
+ ⚹
_____

∧
+ ꓵ∧
_____

Lah'aj Kids

ADDITION WORKSHEETS WITH ŠABATIYYA

ᒡᗷᒡᗷ1·RANAN·NAME

ⵝᕼᐃᒣⵁ·KHADUT·DATE

Lah'aj Kids

ADDITION WORKSHEETS WITH ṠABAṪIYYA

ᚦᚾᚦᚾᚦᛏ · RANAN · NAME

/⌂⌂✕ · 25

ᚷᚻᚦᚼᛉ · KHADUT · DATE

ᚻᚦᚾ ᚻᚦᚾ ᚹᚦᚳᚷᚦᚷ · SAHAF ASHU ATHU · PAGE 12

Lah'aj Kids
ADDITION WORKSHEETS WITH ŚABAṪIYYA

ᚠᚾᚦᚾ1 · RANAN · NAME

ᚦᚺᚦᚾᚦ · KHADUT · DATE

ᚱᚨᚾᚨᚾ • RANAN • NAME

ᚲᚺᚨᛞᚢᛏ • KHADUT • DATE

/∩∩✳ · 25

ᒣᔕᒣᔕᒣ · RANAN · NAME

/∩∩⋇ · 25

᙭ᕼᗞᕼᔕᘝ · KHADUT · DATE

$$+ \frac{∩^{\curlywedge}}{⋇}$$

$$+ \frac{∩^{\curlywedge}}{I}$$

$$+ \frac{\curlywedge}{⋇}$$

$$+ \frac{I}{\times}$$

$$+ \frac{⊙}{I}$$

$$+ \frac{∩}{⋇}$$

$$+ \frac{I}{\times}$$

$$+ \frac{∩^{\curlywedge}}{∩^{\curlywedge}}$$

$$+ \frac{\curlywedge}{∩}$$

$$+ \frac{∩II}{⋇}$$

$$+ \frac{∩}{∩}$$

$$+ \frac{\curlywedge}{∩}$$

$$+ \frac{⋇}{I}$$

$$+ \frac{\curlywedge}{I}$$

$$+ \frac{∩^{\curlywedge}}{\curlywedge}$$

$$+ \frac{∩^{\curlywedge}}{∩II}$$

$$+ \frac{\curlywedge}{\curlywedge}$$

$$+ \frac{⋇}{⋇}$$

$$+ \frac{I}{⊙}$$

$$+ \frac{⋇}{\curlywedge}$$

$$+ \frac{\curlywedge}{⊛}$$

$$+ \frac{⊙}{\curlywedge}$$

$$+ \frac{I}{I}$$

$$+ \frac{⋇}{⋇}$$

$$+ \frac{∩I}{ }$$

Lah'aj Kids

ADDITION WORKSHEETS WITH ṠABATIYYA

25

RANAN · NAME

KHADUT · DATE

Lah'aj Kids

ADDITION WORKSHEETS WITH ŚABATIYYA

/ΠΠ✕·25

ᚱᚨᚾᚨᚾ·RANAN·NAME

ᚷᚺᚨᚾᚢᛏ·KHADUT·DATE

✳ + ⋏	∩Λ + ⋏	✳ + ⊕	∩Λ + ✳	I + I✳
I✳ + ∩	✳ + ⋏	Λ + ∩Λ	⋏ + ⋏	⊕ + ∩
∩Λ + ∩	⋏ + Λ	∩ + ⋏	Λ + I	✕ + ⊕
◎ + ◎	I + ∩	⋏ + ✚	✳ + ✳✳	✕ + ∩Λ
∩ + ∩	✳ + ✳✳	✳ + ✳✳	∩ + ✚	∩✳ + ∩✳

Laha'j Kids

ADDITION WORKSHEETS WITH ŚABATIYYA

· 25

RANAN · NAME

KHADUT · DATE

Lah'aj Kids

ᒧᗜᒪᗜᒌ·RANAN·NAME

᙭ᕼᗜᓂᗘᖻ·KHADUT·DATE

Lah'aj Kids

/ⵔⵔჯ·25

⊏⌂⊓⌂⊏1·RANAN·NAME

ჯ⊞⌂⊼Ψ·KHADUT·DATE

⊙	✳	┃	┃	∧
+ ⊕	+ ✳	+ ✳	+ +	+ ⊙

✳	∩∧	⊕	✳	∧
+ ∩∧	+ ⊙	+ ⊕	+ ✳	+ +

┃	✕	✳	∩┃	+
+ ⊙	+ ∩┃	+ ✳	+ ∧	+ ✳

∩┃	✳	✕	✳	∩∧
+ ∩┃	+ ⊙	+ ┃	+ +	+ ┃

⊙	✳	∩∧	+	∩┃
+ ✕	+ +	+ ∧	+ ✳	+ ┃

/∩∩⋇·25

ᒋᐢᒉᐢᒋ·RANAN·NAME

⋇ᕮᐃᕊᡃ·KHADUT·DATE

Laháj Kids

ADDITION WORKSHEETS WITH ŚABATIYYA

/25

RANAN · NAME

KHADUT · DATE

ᖶᖵᖶᖔ1·RANAN·NAME

ᕽᖁᖜᖔᕿ·KHADUT·DATE

Lah'aj Kids

ADDITION WORKSHEETS WITH ṢABATIYYA

ꓒꔠꓒꔠꓔ·RANAN·NAME

ᚷⴼ△ꔠᚹ·KHADUT·DATE

$$+ \begin{array}{c} \curlywedge \\ \cap \sqcup \end{array}$$
$$+ \begin{array}{c} \circledast \\ \cap \sqcup \end{array}$$
$$+ \begin{array}{c} \times \\ + \end{array}$$
$$+ \begin{array}{c} \odot \\ \cap \end{array}$$
$$+ \begin{array}{c} \dagger \\ \mid \end{array}$$

$$+ \begin{array}{c} \circledast \\ \times \end{array}$$
$$+ \begin{array}{c} \divideontimes \\ \odot \end{array}$$
$$+ \begin{array}{c} \cap \\ \divideontimes \end{array}$$
$$+ \begin{array}{c} \circledast \\ + \end{array}$$
$$+ \begin{array}{c} \wedge \\ \circledast \end{array}$$

$$+ \begin{array}{c} \mid \\ \cap \wedge \end{array}$$
$$+ \begin{array}{c} \times \\ \divideontimes \end{array}$$
$$+ \begin{array}{c} \divideontimes \\ \mid \end{array}$$
$$+ \begin{array}{c} \cap \\ \wedge \end{array}$$
$$+ \begin{array}{c} \mid \\ \cap \sqcup \end{array}$$

$$+ \begin{array}{c} \mid \\ \odot \end{array}$$
$$+ \begin{array}{c} \dagger \\ \curlywedge \end{array}$$
$$+ \begin{array}{c} \divideontimes \\ + \end{array}$$
$$+ \begin{array}{c} \cap \\ \divideontimes \end{array}$$
$$+ \begin{array}{c} \odot \\ \divideontimes \end{array}$$

$$+ \begin{array}{c} \odot \\ \circledast \end{array}$$
$$+ \begin{array}{c} \cap \sqcup \\ + \end{array}$$
$$+ \begin{array}{c} \dagger \\ \cap \end{array}$$
$$+ \begin{array}{c} \wedge \\ + \end{array}$$
$$+ \begin{array}{c} \cap \sqcup \\ \circledast \end{array}$$

Lah'aj Kids

NOTES

NOTES

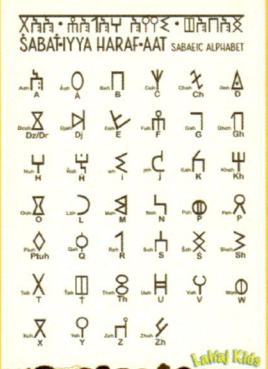
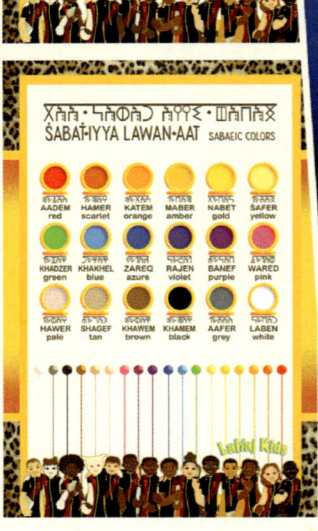